Bugs

and Butterflies

WRITTEN BY JANE MANNERS

ILLUSTRATED BY DONNA REYNOLDS

Scott Foresman Addison Wesley

http://www.sf.aw.com

CHILDREN'S TELEVISION WORKSHOP

http://www.ctw.org

ISBN: 0-201-31744-3

Printed in the United States of America.

2 3 4 5 6 7 8 9 10-BX-01 00 99 98

Annie had a garden filled with pretty flowers and yummy vegetables.

In the garden, bugs, slugs, and butterflies crept and crawled and flew.

"Annie," said Herman, "let's count all the little creatures."

Annie and Herman started with the slugs.

Annie counted 4 slugs on the vegetable side of the garden. Herman counted the same number of slugs on the flower side.

“That’s 4 slugs here and 4 slugs there,” said Herman. “How many is that in all?”

"4 and 4. Count them," Annie said. "There are 8 slugs in all."

Next, Annie and Herman counted the ladybugs.

"I see 4 ladybugs," Annie said.

"3 ladybugs over here!" Herman called out.

"4 and 3. Count them," Annie said. "That's 7 ladybugs in all."

"Oh, no," said Annie. "My 4 ladybugs are flying away!"

Annie and Herman waved good-bye to the ladybugs.

“That means we won’t have 7 ladybugs anymore,” said Annie. “4 flew away. How many are left?”

"There are 3 ladybugs left," said Herman.

Finally, Annie and Herman counted the butterflies.

"I see 3 butterflies," Annie said.

"And I see 2 butterflies on my side," Herman said. "3 and 2. Count them. That's 5 butterflies in all."

"Oh, no, Annie!" cried Herman. "2 of your butterflies just flew over to my side of the garden!"

“Now how many butterflies are there?” Annie wondered.

"Now I have 1 butterfly, and you have 4 butterflies," Annie said.

"1 and 4. Count them. There are still 5 butterflies in all," said Herman. "It's the same number!"

Annie smiled. "That's right!"

2 more ladybugs flew into the garden.

"We will never count these bugs and butterflies," cried Annie. "They come and go too fast."

Herman saw the slugs. They moved slowly.

"Annie, I know how many creatures are in our garden today," Herman said. "Lots of butterflies, lots of ladybugs, and exactly 8 slow slugs!"